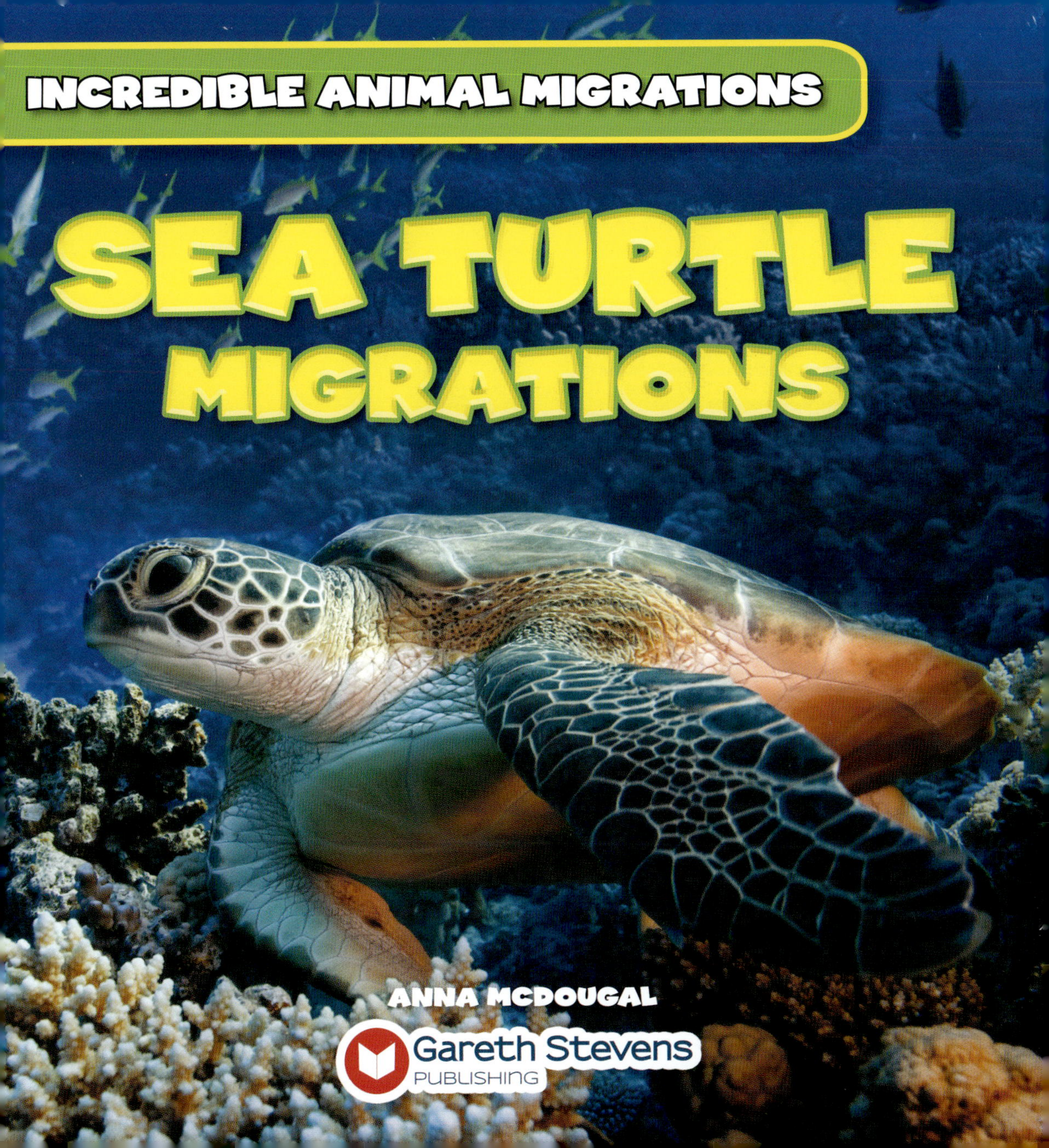

INCREDIBLE ANIMAL MIGRATIONS

# SEA TURTLE MIGRATIONS

ANNA MCDOUGAL

Gareth Stevens PUBLISHING

**Please visit our website, www.garethstevens.com. For a free color catalog of all our high-quality books, call toll free 1-800-542-2595 or fax 1-877-542-2596.**

**Cataloging-in-Publication Data**
Library of Congress Cataloging-in-Publication Data
Names: McDougal, Anna, author.
Title: Sea turtle migrations / Anna McDougal.
Description: New York : Gareth Stevens Publishing, [2023] | Series: Incredible animal migrations | Includes index.
Identifiers: LCCN 2022011202 (print) | LCCN 2022011203 (ebook) | ISBN 9781538278376 (library binding) | ISBN 9781538278352 (paperback) | ISBN 9781538278369 (set) | ISBN 9781538278383 (ebook)
Subjects: LCSH: Sea turtles–Migration–Juvenile literature.
Classification: LCC QL666.C536 M395 2023 (print) | LCC QL666.C536 (ebook) | DDC 597.92/8–dc23/eng/20220321
LC record available at https://lccn.loc.gov/2022011202
LC ebook record available at https://lccn.loc.gov/2022011203

Published in 2023 by
**Gareth Stevens Publishing**
111 East 14th Street, Suite 349
New York, NY 10003



Designer: Rachel Rising
Editor: Caitie McAneney

Photo credits: Cover, pp. 1, 19 Willyam Bradberry/Shutterstock.com; pp. 3, 4, 6, 8, 10, 12, 14, 16, 18, 20, 22-24 moomsabuy/Shutterstock.com; p. 5 Drew McArthur/Shutterstock.com; p. 5 Shane Myers Photography/Shutterstock.com; pp. 7, 13 Manuel Ocen/Shutterstock.com; p. 9 NaturePicsFilms/Shutterstock.com; p. 11 Yana Georgieva/Shutterstock.com; p. 15 Subphoto.com/Shutterstock.com; p. 17 StacieStauffSmith Photos/Shutterstock.com; p. 21 optimarc/Shutterstock.com.



Printed in the United States of America



CPSIA compliance information: Batch #CSGS23: For further information contact Gareth Stevens, New York, New York at 1-800-542-2595.

# CONTENTS



**Boldface** words appear in the glossary.

## What's a Sea Turtle?

Sea turtles spend most of their lives in the ocean. They can be found in warm and cool waters around the world. There are seven different kinds of sea turtle. Two kinds of sea turtles are loggerheads and green sea turtles.

GREEN SEA
TURTLE

LOGGERHEAD
SEA TURTLE

## Turtles on the Move!

Most sea turtles migrate, or move around seasonally. They are born on land. Next, they swim to the open ocean. Then, they return to the place they were born to lay eggs. Leatherback sea turtles make the longest trip. Loggerheads migrate a long way too.

# Meet the Loggerhead Turtle

Loggerheads are sea turtles with large heads. Their heads have strong jaws. That helps them eat sea animals with hard shells. Loggerheads are found around the world. Many loggerheads live in waters off the **coast** of the United States.

Loggerheads mostly eat meat. They often eat things that float. That's why ocean trash can harm loggerheads. They spend much of their life in the open ocean. Loggerheads have to come to the water's surface to take in air. They can hold their breath for a long time though.

## Baby Turtles

Mother turtles find a beach to nest on. They lay eggs in the sand. People need to leave these eggs alone. Soon, they **hatch**. Baby sea turtles come out of the eggs. They make their way to the open ocean.

## The Open Ocean

Baby turtles are often meals for bigger ocean animals. However, some are able to grow up. They spend about 7 to 15 years in the open ocean. The open ocean is very deep. Sea turtles **forage** for food near the ocean's **surface**.

## Back to the Coast

After a few years, it's time to migrate! Loggerhead sea turtles move to coastal waters. When they're ready, mother sea turtles will go on land to lay eggs. This is often close to where they were born.

SOUTH CAROLINA
DEPARTMENT OF NATURAL RESOURCES
LOGGERHEAD TURTLE NESTING AREA
Eggs, Hatchlings, Adults, and Carcasses are Protected By Federal & State Laws

## A Long Trip

Loggerhead migration paths are different. That's because they live in different areas of the world. Pacific Ocean loggerheads nest in Japan and Australia. Then, they migrate across the ocean to California, Mexico, and South America. That's almost 8,000 miles (12,875 km)!

## Awesome Ocean Animals

Migration is an **adaptation**. It helps sea turles live in their ocean home. You can help sea turtles keep migrating for years to come! If you see sea turtle eggs on a beach, leave them alone. If you see trash near the water, pick it up.

# LOGGERHEAD SEA TURTLE ADAPTATIONS

- migrating for food and nesting
- can hold breath for hours
- strong jaws for eating shelled animals
- strong flippers for swimming

# GLOSSARY

**adaptation:** A change in a type of animal that makes it better able to live in its surroundings.

**coast:** Land next to the ocean.

**forage:** To go from place to place looking for food.

**hatch:** To break open or come out of.

**surface:** The top part of something.

# FOR MORE INFORMATION

## BOOKS


Esbaum, Jill. *Sea Turtles.* Washington, DC: National Geographic, 2021.

Moore, Lindsay. *Yoshi and the Ocean: A Sea Turtle's Incredible Journey Home.* New York, NY: Greenwillow Books, 2022.


## WEBSITES

**Green Sea Turtle**
*animalfactguide.com/animal-facts/green-turtle/*
Learn more about the green sea turtle and its incredible migration.

**Loggerhead Turtle Facts!**
*www.natgeokids.com/uk/discover/animals/sea-life/loggerhead-turtle-facts/*
Explore more fun facts about loggerhead sea turtles with National Geographic Kids.

**Publisher's note to educators and parents:** Our editors have carefully reviewed these websites to ensure that they are suitable for students. Many websites change frequently, however, and we cannot guarantee that a site's future contents will continue to meet our high standards of quality and educational value. Be advised that students should be closely supervised whenever they access the internet.

# INDEX